Restoration

OF A

NEW ENGLAND FARM

The Booth-Dimock Homestead

Restoration

OF A

NEW ENGLAND FARM

The Booth-Dimock Homestead

Susan R. Crossen
and Thomas J. Crossen, Jr.

TWO HARBORS PRESS

212 3RD AVENUE NORTH, SUITE 290

MINNEAPOLIS, MN 55401

612.455.2293

WWW.TWOHARBORSPRESS.COM

ISBN-13: 978-1-934938-61-4

LCCN: 2010941887

DISTRIBUTED BY ITASCA BOOKS

COVER DESIGN AND TYPESET BY JAMES ARNESON

PRINTED IN THE UNITED STATES OF AMERICA

To all of the previous owners through the generations who cared for and loved the Booth-Dimock farm.

Contents

Foreword

Susan R. Bergeron and Thomas Crossen, Jr., met in 1969 at H.H. Ellis Technical School in Danielson, Connecticut. Although neither one would end up with careers in the particular trades studied there, it was the beginning of their lives together. They were married on August 1, 1970, at the ripe old ages of seventeen and eighteen. Shortly thereafter they started their family. Susan remained at home to care for their two children. Tom worked a day job as an apprentice in the electrical field and a part-time job as a plumber's assistant. He also attended classes for his electrical license in the evenings.

Tom's exposure to the construction field ultimately led the couple in the direction of renovating houses. Both Tom and Susan did the work while living in the homes. This included cleaning, painting, laying carpet and vinyl flooring, hanging sheetrock, etc.; basically, whatever improvements were deemed necessary.

After remodeling a dozen or so projects and selling them for a small profit, Tom decided it was time to transition into the residential construction business. He first started building modular homes, and then went on to the more traditional stick-built house. Susan was enlisted to decorate and furnish the model homes built on speculation in their many subdivisions.

From there the company, Crossen Builders, Inc., grew to be a substantial small business, producing at its height over one hundred and twenty-five homes per year. As the business evolved, the construction of apartments, condominiums, and small commercial properties were added to their portfolio of accomplishments.

After many moves during the remodeling years, the Crossens finally settled in the town of Tolland, Connecticut, in 1985. The fifty-two acre parcel of land, upon which they constructed a colonial house, became the place they called home. The Crossens also reconfigured the existing dairy barn into a stable for horses, and they built an indoor riding arena. Tom and Susan single-handedly returned the overgrown fields back to useable pasture and installed board fencing for their horses.

Quite a few years later, a house fronting some of their back acreage further down the street came up for purchase. At the time, they did not know that this was a historic house. They would later learn that it was the Deacon Francis West House built in 1721. The Crossens purchased the house as an investment and an exit route for a road in the event they ever decided to subdivide their farm. It would be their first attempt at a restoration, a far cry from their earlier renovation projects. This is where they would get their feet wet before taking on their largest restoration project: the Booth-Dimock property located in Coventry, Connecticut.

Timothy K. Gill

First Impression

*I*n the spring of 2001, my husband Tom came home all enthused about a piece of real estate that caught his attention. He insisted on taking me for a drive by the farm located in a neighboring town. My first impression of the property wasn't nearly as positive as his. In fact, it was quite the opposite. I surmised rather quickly that it would require a tremendous amount of work and resources to restore the property to its original grandeur.

The colonial house was sided with cedar shingles that were turning black. Part of the stone foundation was covered with black plastic that was flapping in the wind. An enclosed entryway with a squat roof and mismatched windows extended out from the front of the house. The side porch appeared to be constructed in the 1950s, replacing whatever had been original to the house. The windows, six panes over a large single pane, were replacements from around the same time period. On the gable ends of the house, there were only two windows, one above the other, instead of the usual four. The storm windows were aluminum framed. The light gray asphalt shingles looked worn down to a flat white.

**House prior
to restoration
(above)**

**House and side
porch**

**North view
of Barns**

The broken-up gravel and asphalt driveway was wider than normal, giving it a commercial appearance. There were only two large trees and a scant lawn for landscaping in the house common area. Fallen-down stonewalls and debris were in close proximity to the house. Despite all of this, there was a hint of the house's antique origins evidenced by the dentil molding near the roofline and at the gable ends inside the pediments. Still, the overall look was that of an out-dated New England farmhouse in need of major renovations.

From the exterior, the barns looked as though they were nearly about to collapse. The siding, what was left of it, had all but fallen off. The large barn had water damage and was bowing out in the mid section along its length. There was additional water damage in the back of the barn where the two barns conjoined. The milk parlor, which had been an addition, had fallen in onto itself. The roof was a rusty corrugated metal commonly seen on old dairy barns. The sheds near the barns were falling in and down. The barns had only a few windows that were either broken or boarded up.

The land was overgrown with a type of mulberry rose bush, allegedly recommended by the government years prior to fence in cows. Unfortunately, it spread like a weed, even into the woodlands. In addition, the fields were plowed and then left in that condition with ruts throughout the upper tillable land.

Tom and I drove up and down Bunker Hill Road amazed at the amount of road frontage. The large parcel of land would lend itself easily to a subdivision. But, I am ahead of myself and how the sequence of events unfolds.

Barn front with bow

Fields overgrown with brush

The Acquisition

*T*om drove from his office in Tolland, an abutting town, to check out a piece of property located on Flanders Road in Coventry, Connecticut. What a surprise to discover it was his grandfather's farm that he remembered visiting as a child! It was a nice piece of real estate, but not one that would suit his development criteria. On the way, he drove past another farm, which had a fantastic view and substantial road frontage. It was the epitome of a classic New England setting with its house and barns. He thought, *If that place were to ever come on the market. . . .* Barely a month went by when a real estate agent contacted him about a farm in Coventry that had just been listed for sale. When Tom realized it was the place he had driven by, he immediately called the agent back and made an appointment to inspect the property. Tom said, "The barn interiors were all I needed to see." He signed the contract with his agent, who suggested that perhaps Tom should make an offer. Tom instructed him to write the contract for the full amount, suspecting the property would end up in a bidding war. Sure enough, late in the day, Tom was informed that another party was interested and had offered more than the asking price. Tom made his highest and best offer through his agent to purchase the farm. We found out sometime later that our winning offer was only slightly higher than the competing offer.

Initially, we thought we purchased the place for a reasonable price. Since the farm consisted of ninety-two acres with over a mile of frontage, how could one go wrong?

Had we gone the usual course and developed the property, it would have been a much easier and more lucrative undertaking. But as fate would have it, once we started to investigate the history of the property, we found ourselves drawn toward the restoration aspect instead.

The Initial Inspection
of the House Interior

*M*y first view of the house interior was done after the closing. It reflected the Fifties farmhouse style of the exterior. We entered through the kitchen, which at that time was at the end of the house closest to the barns. There were dark stained cabinets on one wall that included an out-dated electric stove. Hidden behind the stove was the face of an original double-sided fireplace. The flooring was neutral-colored linoleum. The wood trim was painted white, as were most of the doors throughout the house. In the breakfast nook area, the walls were painted a light aqua and had dark wainscoting covering the lower section.

Wall with fireplace hidden behind kitchen cabinets

To get to the next room—which spanned the entire width of the back section of the house—one would have to fit through a "hallway" between the huge fireplace with beehive oven and the outside wall. The space was only about two feet wide. Upon entering this room a large fireplace veneered with a Victorian face of hand-cut stone was the focal point. The lintel stone was cracked. There was a chimney cupboard and a small door alongside the fireplace. Behind this narrow door was a set of original stairs that led to the second floor and from there up to the attic. These stairs were made of quarter-sawn oak and they were very steep.

Victorian fireplace veneer

Back set of stairs

The treads were so narrow that one's foot extended past the end of the step. It was suggested that originally the household work staff would have used these stairs. Going back down through the wide room we noticed the floors were uneven and sloped to the left. The walls on half of the room, which had been used as a dining area, were wallpapered in a pattern likely to be found in a farmhouse kitchen. The opposite end of the room functioned as a den and was finished with full-length, manufactured wood paneling.

This area also had some wainscoting on the walls; it was stained a dark walnut. The trim surrounding both of the windows and doorframes, added sometime along the line, was stained the same dark walnut. Just off the den was a full bath with an exposed cast-iron pipe running from floor to ceiling. The walls were covered with aqua tiles to mid-height and a black border ran along the top edge.

Area at end of den/dining area

The hallway leading to the formal parlor was directly in line with the large fireplace and ran toward the front of the house. Off of this foyer there were two front rooms, thought to have originally been the formal parlors. In both of these front rooms there were fireplaces closed off with insulation and boards. After removing these items to inspect the interiors, we found loose bricks that had fallen inside the firebox from the

Opposite end of dining area

Den area partially stripped (above)

Foyer, back view (left)

chimney. In one of the rooms, most of the fireplace trim and the chair rail were missing. Luckily, the opposite front room had these original details, which we could use to replicate the missing pieces. The wallpaper patterns found in each of these front rooms and throughout the house were in the mid-twentieth-century motif.

Fireplace as found in front bedroom (left)

Original front door revealed (below)

The formal front door, which had a few inches cut off the bottom, was still spectacular.

It had the original fan-type window above with wood muntins still intact. The exterior doorframe still had the original grooved trim boards making up the pilaster. The enclosure added onto the front of the house saved the original formal front door and its surrounding trim.

The interior staircase leading to the second floor was narrow and steep in design. The steps and handrail were painted white. The original plaster walls were painted aqua.

Upon arriving at the landing, we noticed strip wood flooring over the top of what we would later find to be wide-board maple. In the first front bedroom there was more strip hard wood flooring, underneath which we unearthed hand-planed tulipwood. Under the strip flooring in the other front room, we found original oak flooring. Tom and I later had these two front room floors painted and glazed with a medium stain. The rest of the restored flooring throughout the house was stained a warm antique brown.

Going straight through from the top of the stairway to the center of the house was a great room that looked as though it had

Original floors uncovered

been a large bedroom or a parlor for an apartment. It had a large piece of linoleum on top of the original pine floors. The ceiling was covered with white paneled acoustical tiles. Off to the right of the center room was a small bedroom with two door openings. This room would later become the main bathroom upstairs.

Alongside a fireplace chimney in the center room there was a hallway that led to a back apartment with another kitchen. Off of that room there was a full bath with a maroon wall-mounted sink and a white cast-iron tub. A set of stairs off the back of

Front staircase before

the house led to the ground floor. To the right of the large fireplace chimney were steps up to the attic. The attic was missing most of its floorboards—there were only a few to walk on in limited areas. There was creosote leakage evidenced by amber fluids long dried all over the exterior of the chimney brick. We determined it was

Creosote covered chimney stack

a miracle the house never burned down. While in the attic, we were ecstatic to find an original period window sash! We would later use it as the template for making the twelve-over-twelve window replacements.

Upon initial inspection the basic construction of the house seemed acceptable, though there were the usual uneven floors and crooked doorways common in older homes. A termite inspector checked the wood beams in the basement. He assured us that whatever damages had been done were in the past. There were some dried out beams but nothing too worrisome looking; at least not from his vantage point.

Masonry & Stone Foundation Repairs

Chapter 4

*W*e proceeded as one normally would, with the basement being the first space to address. Since we were going to extend the house ten feet and add a carriage shed/garage to the gable end of the T, we were going to need a tremendous amount of stone. Luckily, at the same time, Tom was doing another job in the neighboring town of Windham, where his excavator had uncovered a shale type ledge of flat stones. This was perfect for use in the basement and carriage shed floors. Truckloads after truckload of stones were unloaded onto our field, easily covering over an acre of land. Additionally, we were going to need a different kind of finished granite stone for the exposed part of the house and carriage-shed foundation. We were fortunate in that regard as well; behind the large barn was a stone wall made of hand-cut stone that matched the house foundation perfectly.

Our luck was soon to run out. After assessing the foundation, it was determined that we needed to replace an existing section—a crawl space with inadequate frost protection—with a full basement to the outside wall. The work began by supporting the house with carrying posts and removing the

Rebuilding fireplace in front palor

unsupported rock foundation. Then the dirt in the crawl space area was removed. At this point we were able to determine that the wood framing above was originally a summer porch added onto the north side of the house. It had collapsed downward having had no real support from the beginning; thus, the reason the floors on the first floor had a slope to them on that side of the house.

Next, we met with the stonemasons whom we had found six months earlier working on the Daniel Benton Homestead Museum in Tolland, Connecticut. They needed to finish that job before coming to appraise

ours. It was the beginning of a long-term relationship. We can't say enough about the work ethic of Ray Paine and Steve Breen. They ended up working on our project full-time for three years. Upon completion of their evaluation it was determined, to our dismay, that all of the fireplaces needed to be rebuilt from the first floor through to the chimney tops. The stone bases in the cellar were structurally sound but the brick fireplaces, which had originally been constructed with a clay mortar, had sustained water damage. This caused the mortar to deteriorate, which loosened the bricks and ultimately rendered the fireplaces unusable. We kept most of the original bricks to reuse them where they would be most visible. The masons were able to acquire a similar small, reddish-orange brick salvaged from an old mill to make up the difference.

Since we wanted everything to be done as authentically as possible, the fireplaces were rebuilt in the tradition-al manner. This basically meant that we would use brick in the entire structure. No cinder blocks in the hid-den areas or masonry fillers were used. The only non-period additions would be the flues and dampers. This made the fin-ished fireplace approximately three inches wider than the original footprint. To accom-

Beehive oven and fireplace uncovered

modate this change, the original mantel and trim had to be made that much wider. There is a small line in the trim where that addition can be seen on the finish face trim board. The masons also went to the trou-ble of saving the original clay mortar and then using it for the firebox, which created an authentic look that is rarely achieved. Most often modern mortar is dyed to imi-tate period colors.

The large stone fireplace needed a tre-mendous amount of work. First, the "up-dated" Victorian front of cut stones was re-moved; then the cracked lintel stone was replaced. This required a cinch-type pulley

Beehive Oven

system to lift the heavy stone out and put in a replacement. Some of the other granite stones needed to be replaced as well. The fireplace was opened from the first to second floors. A huge stone that separated the chimney boxes for the double-sided fireplace was hoisted up into the neck of the fireplace and through to the second floor. Brick was used from that point up to the chimney top. The nearby beehive was no longer safe and needed to be entirely rebuilt. This was accomplished by building a mound of sand in the shape of a beehive oven. Bricks were then laid on top of the sand mold and mortared together with three layers of brick. After the mortar dried, the sand was removed by digging it out with a trowel. A large stone nearly three feet wide and eight feet long was acquired for the hearth.

Since we needed to rebuild all of the fireplaces from the first floor up, we decided to add a fireplace in the north front upstairs bedroom and a second fireplace in the upstairs master bedroom. We also needed to replace some of the lintels on the other existing smaller fireplaces. The hearths for all of these would be granite stones hand-cut to the correct size. The house now had a total of seven working fireplaces plus the beehive oven.

Back in the basement, the masons started the tedious job of hand-building a fieldstone foundation to match the existing one. The fieldstones were obtained from a stone wall that was conveniently located near the north side of the house. The top of the foundation was eventually finished with the formal cut granite stones. It took months to complete the thirty-nine-foot foundation wall. Afterward, a black sealant was applied to the exterior to keep water from leaking into the basement.

North wall under construction

North view of house under repair

hand-cut granite stones were used. The finished wall spanned more than two and a half feet in depth. While the gable end was still open, wall-to-wall flat stones were installed for the basement floor. A skid-steer was required to move the heavy stone. Prior to that, drainpipes had been placed beneath the surface to make sure the basement remained dry. This additional work took the labor crew another two months to complete. And this had to be finished before the work on the carriage shed could begin.

The back gable end of the house and the attached carriage shed also required hand-laid foundations. Support beams were used to hold up the end of the house while the replacement stone foundation was built. On the interior side, fieldstones matching the rest of the basement were used. On the reverse carriage shed side,

Replacing the sill

Basement original foundation

Basement with new hand-laid foundation

The Carriage Shed

*W*hen the time came to start the carriage shed, the concrete footings for the stone foundation walls were poured first. Then, the foundation walls were built using the cut granite stones found in the stone wall behind the barns. The finished carriage shed stone foundation matches the barns' foundation perfectly; even the mortar matches in both color and application.

Earlier, we searched for materials to construct a garage that would ultimately look like an antique carriage shed. We contacted an antique collector who was recommended to us. There was a possibility that he might have the materials we needed. He happened to own the Gillette House formerly located on Main Street in Glastonbury, Connecticut. It had been dismantled piece-by-piece in 1988 and stored in a tractor-trailer ever since. We purchased the disassembled house and had it delivered in the box trailer. The pieces were placed on open ground to determine how it could be reassembled to fit the size of the awaiting foundation. The building crew only had a rendering of what the carriage shed should look like. A full set of plans could not be drawn in advance, as the exact ma-

Gable ends of conjoined barns & footings for carriage shed

terials of the post-and-beam house were unknown. After much consideration, measurements, and drawings, the master carpenters were able to assemble the garage by cutting down some of the larger beams and smaller joists. The antique wood board sheathing was procured elsewhere but looked as though it could have been original to the carriage shed. Lastly, we duplicated the stonework that was done in the house basement for the flooring.

Carriage shed in framing stage

The footprint of the house and carriage shed appeared the same as the original, as shown in a photograph dated 1926.

Finished interior of carriage shed (above)

Picture of farm in 1926 taken by Walter Thorp (right)

Major Problems Arise

*A*s the carriage shed was constructed, new problems were being uncovered in the main house. Upon removing the floorboards to replace them with reproduction old-growth wide pine flooring, our framing contractors discovered dry rot in the massive sill beams and many of the floor joists. Unfortunately this was not detected from the basement vantage point, as the problems existed on the top of the joists—which were not visible from below. We hired a woodcutter to cut replacement lumber from our own trees. He used a portable sawmill to reproduce the floor joists that would be visible from the basement. He also made the underlayment for the wood floors by cutting a tree *through and through* (width of the board was the true width of the tree). The boards were wider at the base and narrower at the top. The bark was scraped off the board's edges and installed over the floor joists. Now, instead of seeing plywood underlayment from the basement, these natural boards are featured. We eventually stained all of these underlayment boards, floor joists, and sill beams brown to give them an antique appearance.

Yet another major disconcerting problem was uncovered when removing the finish trim on the cor-

Floor joists revealed (above)

ner of the two front rooms. All exterior posts that ran from the sill to the roof-line were completely rotted—most likely destroyed by termites. These needed to be replaced along with the sills on top of the stone foundation wall. This was a formidable task, as each twenty-foot-high post, eight inches square, would be installed from the exterior one at the time. They were mortised into the beams on the first and second floors, and then tied into the attic. To add further insult to injury, when the addition of the second floor was done in the back section of the house sometime in the mid-nineteenth century, the builder did a poor job connecting it to the front section of the house. This is where the large center room above the kitchen was located. All of this required procuring and replacing more huge beams.

Then another blow: while attempting to remove shingles from the roof of the back

Floor joists replaced (below)

Scraping bark off wood for subflooring

the culprit and the entire roof had to be replaced—both frame and sheathing. This required the return of the woodcutter to make exact replicas of the original roofing timbers. These were very long roof rafters that would run from the roof peak to the top of the sidewalls. After the roof was repaired, pressure-treated Southern yellow pine shingles were applied. The life expectancy of these is similar to the old-growth cedar originally used. Needless to say, the cost to restore this house was increasing by the minute. But, it was too late to turn back at this point. The time and money already spent on the foundation, fireplaces, the entire replacement of the first floor joists and beams, and the work underway on the carriage shed, made it impossible to abandon the project.

section of the house, the roofing crew reported that the underlying boards were coming off with the shingles. Dry rot was

It was as though the house had revealed its imperfections, a little at a time, to make it impossible for us to retreat.

Decisions & Procurements

*W*hile the aforementioned work progressed, the ever-changing floor plans were also taking shape. One of the biggest decisions was to move the kitchen to the center of the house where the rebuilt stone fireplace with brick beehive oven was located. This was certainly the location of the original kitchen. We also deduced that if we moved the back stairwell to the north side of the house, we'd be able to fit a pantry off the kitchen. This would at the same time create a standard-width hallway in the center of the house, replacing the existing narrow space along the sidewall. In that area alongside both fireplaces, we restored the original cupboards and added chimney closets.

In the meantime, the interior walls were opened up to install the insulation and mechanicals, including new wiring, plumbing, heating and air condition-ing, a security alarm, and even a central vacuum system. The framing for the window openings needed to be reduced back to their original size. This would accommodate the custom handmade, mortise-and-tenon, twelve-over-twelve windows with antique reclaimed glass. These were made of

Eastern white pine, determined to be the same material used in the originals.

During the natural progression of the job, many decisions had to be made and products had to be purchased in advance to avoid delays in the project. We attended a restoration convention for the trades in Boston, Massachusetts, and a historic home show in King of Prussia, Pennsylvania. Both shows exposed us to a host of suppliers, subcontractors, manufacturers, vendors, etc. We also referenced many books for ideas and to find products needed for our restoration project.

At the Vermont Soapstone Co., in Perkinsville, Vermont, we acquired a perfect soapstone sink. We also ordered a nicely grained slab of soapstone for the counters. At a later date, we sent the company a plywood template for the cabinet bases. They in turn brought the materials to our jobsite, where they cut and installed the soapstone counters.

Another drive to Stoddard, New Hampshire, led us to Carlisle Wide Plank Floors. This place was a little overwhelming, as they offered a huge selection. We decided on old-growth wide pine boards for the less formal rooms and a slightly narrower width with smaller knots for the formal front rooms. We installed these on the first floor, which had no original hardwoods. The second floor had some original floors that were salvageable. We needed to find additional antique flooring for those areas that were not. We revisited the collector in Glastonbury. He had enough painted floorboards for the three remaining rooms. None of the painted floors were sanded flat; instead, we brought them to a painting shop to be chemically stripped. This maintained the hand-planed quality. We ordered handmade, cut nails to install all of the flooring.

A trip to Essex, Connecticut, provided us with an antique dry sink to use in the ground-level bathroom. A short ride to the neighboring town of Willington to look for wood flooring for the attic was also successful. A local reproduction-lighting shop in Marlborough provided most of the fixtures for the house; others were found and purchased in Sturbridge, Massachusetts. We also attended many auctions and antique shows. At an auction near the town of Cooperstown, New York, we purchased another dry sink for the center island and a corner cupboard for the kitchen. From an antique show in Farmington, Connecticut, we brought home a hutch for the kitchen. A second trip to New Hampshire provided us with a huge cast-iron tub that we came across in an architectural salvage yard.

Husband & Wife Dynamics

Chapter 8

I'd like to share some of my thoughts and feelings I had during this project. Originally, I started working on the place with thoughts that we would eventually sell the house and barns with some of the acreage separately from the rest of the property—perhaps subdivide and build a few houses on the rest of the land. At least that was what was running through my mind. It became apparent early on that Tom had a different idea; he wanted us to move there. Tom can be a very convincing person at times, but he was going to have a hard time selling me on this one. I was comfortable in the home we made in the neighboring town. As he tried to win me over to his way of thinking, I came back with assertions of my own. My list of demands grew as the job progressed. But no matter what I wanted, he was amenable to it. As time passed, the more of me that went into the place, the less I resisted. Besides, I spent so much time on the job that it started to feel like a second home. Eventually, it came to the point where my mission was to make our new residence as functional and comfortable as possible.

This was not an easy road, especially with the unforeseen problems that arose. At times the stress was pretty intense and the amount of work that needed to be done was oftentimes overwhelming. It wasn't just the house and barns that required our attention, but the entire property. One can only imagine how exhausting it was to make all of the decisions for a job of this magnitude and procure all of the necessary materials to see it through to its completion—not to mention the cost that increased with every newfound problem.

Complicating things further was the need to find a diverse group of talented craftsmen, subcontractors, suppliers, and workmen that were skilled in this type of work. What made it even more difficult still was the shortage of individuals trained in the traditional methods of construction. Once this carefully chosen cast was ready in full force, it was important to orchestrate the job as efficiently as possible. Even with everyone working diligently on his or her own perspective tasks, the entire restoration from start to final landscaping took five years.

Work on the Interior: Plaster & Trim

*O*ne of the most used materials in the house was the wood for the finish woodwork. We purchased antique wallboards from an antique collector and a salvage yard. We also ordered a tremendous amount of lumber in varying sizes directly from a millwork shop.

The restoration of the wood mantles on the seven fireplaces required a significant amount of wood, as did the trim around all thirty-seven windows and doors, chair rails, and the large exposed ceiling and wall beams. The pantry walls as well as the stone fireplace wall-surround in the kitchen were covered with reclaimed antique wallboards matched to the original wallboards on the opposite side of the room. More antique wood was used in the den and kitchen wainscoting and the trim along the floorboards. We also used antique lumber to build our gorgeous kitchen cabinets. We acquired the wood years prior, thinking we would use it on the Deacon West house in Tolland, Connecticut. It was purportedly 270-year-old Southern yellow pine flooring that came out of a church. It was so thick that when cut in half (through its length), it provided us with double the material for the front of the cupboard doors.

Fireplace trim configuration

In addition to the vast amount of wood required for the trim work, craftsmen spent endless hours milling it into period detail trim, no longer available on the market. For example, on one fireplace mantelshelf alone there are seven edges with differing angles that progress downward in size. The chair rail in the formal rooms consists of four pieces of different trim and the baseboard has a very thin, fine bead on the top edge. The window trim is a series of beaded edges and band molding. The ceiling beams have an even larger beaded edge. The front formal rooms have a very unique and intricate system of beaded boards and band-molding trim on the fireplace faces—it reverses itself at the corner and repeats on the side that projects into the room. Above the mantelshelf there is a plastered area surrounded by a carved rope-style trim. Although these are all very detailed and elaborate, they do not present themselves in a garish manner but in more of an understated elegance.

Many of the thirty-seven original doors were present in the house when we purchased it. It was pointed out to us that the door jam configuration was very rare in that it was made at a specific angle to form a natural stop when the door is closed. What doors were missing we duplicated with antique reclaimed wood. A blacksmith custom made the wrought-iron door latches throughout the house. The door exiting to the carriage shed was veneered with one finish on the inside and a more rustic finish on the garage side.

As most period houses were trimmed before the plaster was installed, we opted to proceed in that manner. This was a major challenge. All of the wood trim needed to be installed and sealed with a primer to protect it. The floors needed to be stained and sealed as well. Additionally, quarter-inch plywood was placed on top of the floors to protect them while plastering. Since a

Kitchen cupboards

tremendous amount of water is used in plaster, all of the woodwork also needed to be covered in plastic. Our plasterers estimated that about one thousand gallons of water was needed just for the sections on the second floor. Before the plastering began, the exposed beam in the kitchen ceiling was installed. Only the kitchen cupboards would be added after the plastering. To make the plaster as authentic as possible, we collected horsehair for three years and it was added to the mix. It turned out that the stonemasons who made all of the prior fireplace repairs and hand-built foundations were also adept at mixing and applying plaster in the traditional three-coat method. They placed planks on each of the front stair treads and carried the mixed plaster to the house in five-gallon pails. If a rest was needed from carrying the heavy load, the pail was set down on top of the plank instead of the stairs. As one can imagine, this was a very cumbersome job. The plastering alone would take a month to apply. It was done in the autumn when drying conditions were favorable.

For the kitchen cupboards, we hired a master craftsman schooled in the almost lost art of mortise-and-tenon joinery. We met with him to discuss what we were looking for in a finished product. We wanted the cabinets to match the dry sink and hutch that we had previously purchased and intended on using in the kitchen. He drew plans to incorporate the freestanding look of the antique pieces. He also gave the doors flat panels to match those pieces as well. We elected to veneer both the dishwasher and refrigerator with the flat panels. The chosen knobs were round to, again, match the old dry sink. The crown molding was designed to blend with the existing antique pieces. Finally, a plate holder, apothecary, and a bank of drawers were included in the plans. We wanted the finish on the cupboards to match the reddish-brown pine antique pieces. For this job, we used our regular painting contractor. We discussed with him what we wanted for a finish. This ended up being quite a challenge, as the wood itself affected the final color. The first try came out too dark—almost walnut; the second too red; and the third, perfect.

The same color experiment happened with the wood wallboards. The first try came out too orange, and the second too dark. We instructed the painter to strip off some of the dark stain. Midway from entirely stripping the stain, we had him stop. He removed enough stain to lighten the walls, but the recessed areas remained darker. This inadvertently created the exact look we wanted. The end result was a mustard-gold color with an over glaze of warm honey brown. I don't know if it could be reproduced!

Work on the House Exterior

$\mathscr{I}$n addition to all of the interior work, the exterior of the house was also being addressed.

We started with a new open porch on the southeast side of the house. The nine-foot-wide structure was floored with a product called *iron woods*, which was imported from Brazil. Five Roman Ionic pillars made from mahogany and cedar were erected to hold the roof, which in turn was topped with a zinc-coated copper panel system. The panels were eleven by fifteen inches, locked together in a tongue-and-groove system that required each panel to be soldered together with heat from a blowtorch. The biggest challenge was making the hidden built-in gutter system that allows the water to run off the virtually flat roof, preventing water from dripping over the perimeter edge. An exterior-mounted gutter would not be in keeping with a period house. The entire roofing process took much longer than originally estimated. The back of the porch wall was sided with wide, flat boards in varying widths. This was to replicate the original siding in that area as evidenced by some of the boards that were uncovered. The ceiling was covered with flat boards in random widths with a very fine bead.

We then moved on to the task of stripping cedar shingles and the clapboard siding beneath. Our siding contractor discovered a rare type of sheathing method under the original clapboard siding. The horizontal edges of the hand-planed boards had a beveled joint, which made the outside finish flush (instead of the more common butt joint). This was probably done to keep the heavy winds out. Some of the sheathing needed to be replaced at different areas on the house exterior. After that work was completed, the twelve-over-twelve wood windows and storms were installed. Spruce quarter-sawn clapboard siding cut in two-, four-, and six-foot lengths was applied over the sheathing. This was the traditional method used in that time period. Finally, the house exterior was painted an antique white.

The front of the carriage shed was sided with the wide, random-width boards in a similar fashion as the open porch. The rest of the shed was sided with the spruce quarter-sawn clapboard (the same as the house). We used overhead garage doors—veneered to look like traditional wrought-iron, hinged doors—that open up in the middle. We accomplished this look by veneering a standard garage overhead door with a thin coat of cedar boards (run vertically). There is still a crack visible where the door bends between the door panels, but it is not discernable from even a short distance away. We installed hand-made, wrought-iron door handles and hinges just for the effect. Running above both garage doors is a straight transom of windows made with old glass.

Porch

Early on, Tom met with architectural preservationist, William Gould. Mr. Gould possessed a world of knowledge in period construction. He was hired to determine the age of the house based on the materials found in the original construction. This information coincided with the genealogy report done at a later date. In addition to this contribution, he was contracted to draw the designs for the carriage shed, side porch, and front entryway portico. He was responsible for building the custom windows, the French door, and some of the architectural millwork (i.e., trim and dentil moldings that were missing on the exterior of the house). He also repaired the front door and retrofitted it with a period locking mechanism.

At this point in the project, he was working on making handmade reproductions of the dentil blocks along the roofline and in the pediments at the gable ends of the front box of the house. These were assembled using a complex method of joining several

Front portico

different pieces of carved wood that fit to-gether like a jigsaw puzzle.

His next task was to construct and install the formal front portico. He used pillars similar to those used on the side porch, only much more formal—they had *fluting* (grooves) and scrolled and elaborately carved Roman Ionic capitals at the pillar tops. They were situated approximately four feet from the house. The roof was steeply pitched and the underside interior ceiling was arched. The roofline also had dentil molding detail on both the face of the portico as well as the sides. The door was stained to give it the appearance of mahogany—a very period look.

The stone beneath the portico was a gift from my brother, Gerard M. Bergeron, a builder in Worcester, Massachusetts. He acquired it from a remodeling job in the city. We searched everywhere for a stone large enough to act as the base, but were unsuccessful. He was happy to provide us with it as a house-warming present.

We replaced the decorative boards on all of the house corners with twelve-inch-wide boards and ornamental trim, including plinth blocks and crown moldings. Additionally, over the first story windows on the front box of the house, we had window heads cre-ated using formal trim and dentil molding. Each header was covered with lead-coated copper to protect it from the weather.

Interior Decorations

*T*he house interior was the next task of the restoration. This included the wallpaper and paint selections. It took months to make these selections. We purchased books on period styles while on a trip to Sturbridge, Massachusetts. We also toured Sturbridge Village for ideas. I was lucky enough to find two very informative books on the Internet. One book was on window treatments and bed hangings and the other on fabrics used in the Federal period.

One of my best resources came from an antiquary and expert on reproduction wallpaper, Frank Racette. Mr. Racette is renown for providing wallpapers to museums as well as giving lectures on American art history topics at historical societies throughout the Northeast. Frank was very helpful in directing me to the reproduction wallpapers and paints for use in a Federal period house. There was quite a nice selection from which to choose; however, the bold colors and patterns took some getting used to. I spent countless days trying to decide which patterns would flow from room to room. Since many were French toile designs, this was not an easy task. I was immediately drawn to some of the designs and colors, but they didn't always necessarily work in the total scheme. Once one paper was selected, all of the others would need to have some sort of continuity, at least in my contemporary mind. This added many hours to my search. Eventually, I was successful in selecting the patterns for all of the rooms and achieving the desired effect. The paints I chose to match the paper were from companies that offered reproduction, old-fashioned milk-based paints. Most of these happened to work with the wallpaper selections. Those that didn't, we had custom-blended by our very talented painter.

In addition to helping me with the wallpaper selections, Mr. Racette was also instrumental in directing me to the period style of furniture, accessories, and floor coverings that were used in a late Federal home. He even took the time to make up a booklet with those items illustrated for me.

Next on the agenda was the selection of window treatments and bed hangings for the chambers. We were lucky enough to find three antique four-poster Sheraton beds. For the master bedroom, we found a magnificent, flat tester bed that was originally a rope bed made into a double size. The scale of the bed was larger than it appeared in its photographs Online. The wonderful turned posts and blanket rails were more than I could have dreamed. The antique dealer in Virginia from whom we pur-

chased the bed told us that the matriarch of a well-known furniture manufacturer had owned it previously.

At this point, I contacted interior decorator, Gale Horton, who was employed by a local furniture company. I had worked with her twenty years earlier in my previous residence. Tom and I purchased most of our furniture and window treatments through her at that time. The bed hangings and window treatments for this house were a somewhat unique experience for her. I gave her the books I found Online and showed her the photos of the styles I was interested in reproducing. She in turn took all of the measurements for both the windows and beds, designed a pattern for each, and subcontracted the manufacturing. She also used a local seamstress for some of the work. Gale's responsibilities included ordering all of the fabrics and trims, and, ultimately, the installation of the finished products on both the windows and beds.

For the master bedroom, I wanted the bed hangings to be period without hiding the beautiful wood on the bed. The wallpaper is a soft toile in a soothing sage green with a cream background. We purchased the coordinating fabric to be used on both the window treatments and bed hangings. Additionally, the lining fabric and the gold and green tassels and trim needed to be purchased. The desired look was very French.

Master Bedroom

a series of swags with center pieces that resembled ascots or cravats. The trim ran along the bottom edge of the valance and tassels hung from the edge. At the foot of the bed, the valance was the same as on the bedsides but with the addition of a two-foot-long jabot at the two corner posts. The box-pleated bed skirt was fastened to the side rails with heavy-duty sticky tape. The window treatments matched the swag and cravat valance: but instead of a jabot, custom side panels were held back with tie-backs (also trimmed in the gold and green). This room came out quite spectacular, as evidenced by the expression on those who have crossed the threshold.

We then moved on to what we call the "red room." This room is wallpapered in red and cream toile with scenes that depict different seasons. The four-poster cherry bed is a converted rope bed. It is slightly smaller in size than a full bed. This bed was dainty in proportions and has fluting on the foot posts only. The plain posts more commonly associated with this type of bed are used only at the head of the bed. Again, much fabric was needed for the bed hangings, which draped around the sides and were gathered at the headboard and held back with tiebacks. The treatments were not as fancy in this room. There were no gold tassels or trim. Plain piping was used on the pillow shams and coverlet. The gathered bed skirt was made from a diamond-patterned cream and red fabric that coordinated perfectly with the

Master bedroom

There was gold trim used on the bedcover as well as the bedroll. The headboard area had the most fabric, rising from the floor to the canopy top. It was gathered behind the headboard and around to the side, about to the point where the pillows extended on the bed. The bed hangings were then held back with a tieback made of matching gold and green trim. The valance on the tester top, at eighteen inches in depth, was

Red room (above)

wallpaper. The window treatment was totally different than in the master bedroom. Our decorator contacted a museum in Maryland to send additional photographs of the drapes published in the book we were referencing. The museum curator was nice enough to send the diagram of how the effect was achieved using circle hooks and string. There was no valance on the top, and side panels were somewhat gathered into an irregular cascading, downward-ballooning shape created

by pulling the strings up through the small plastic circles.

The next room on the agenda was the children's nursery, which I decorated myself. I was able to achieve a period look by combining the circle-and-string technique used in the red room with some seeded voile sheer panels. The coverlet was plain cotton in a taupe color with a jacquard-style raised print. From a matching king size coverlet, I was able to create the bed skirt

Nursery

in a box-pleated finish. The bed was another four-poster Sheraton without tester that we found in upstate Vermont while looking for fireplace irons. It was stained almost walnut, which I lightly stripped to reveal a dark mahogany. A chest of drawers was purchased with nearly the same color finish. Years prior, I had also obtained a Victorian crib with high posts and a rocker base. With the extra material from the cut-down, king-size coverlet, I made a gathered skirt for the bottom of the crib. I also made a coverlet and matching pillow for it. The antique doll placed inside looks quite comfortable there.

The wallpaper in this room happened to be the most expensive. It was museum quality and hand-pressed, instead of manufactured with industrial rollers. Using seven different colors would require seven different stamps for each section. Again, the pa-

per was a French style, but instead of toile it consisted of animals and flowers called out in French words. The wall covering is stunning with its pink, rose, burgundy, brown, and green colors up against an almost white background. Upon entering this room, our granddaughter, a toddler at the time, became so excited that she broke into a little dance.

The middle bedroom wallpaper is my personal favorite of all. It was a case of love at first sight. This is another Toile de Jouy, but very different from the others. It has a medium dark sage green background with the many cream-colored figures surrounded by a wonderful contrasting burgundy. It has a country scene with figures of people doing everyday work outdoors. One has a young boy astride a donkey lead by his mother. Another scene is of a dog chasing a deer. There are trees with squirrels and a

Favorite wallpaper and French doll

dog closing in on a pheasant. There is another scene of two women picking berries. It is simply a marvelous print. Because this is the bedroom at the top of the formal front set of stairs and visible when one reaches the landing, it was especially important that it make a statement. With this in mind, we procured a Connecticut Valley chest made of cherry wood, stained to imitate mahogany. This transitional piece features a swell front chest with rosewood inlay and flaring French feet. On a popular Internet auction site, we were the winning bidders on a reproduction four-poster, queen-size rice bed with wonderful carvings and fluted posts. We were also able to acquire, from two separate antique dealers, a mahogany English tall chest in the Hepplewhite style and an early twentieth century diminutive Federal secretary in mahogany with satin-wood inlay.

Formal foyer

For the formal front foyer, another toile was selected. This one was designed with gold urns and medallions; green leaves, and muted pink flowers all against an ivory background. The light color of this wallpaper really opened up the area. Prior to that, the foyer was a little on the dark side, having only limited lighting coming in through the transom window above the door. Addition-ally, the wood trim and stairs were painted an antique gold. The stripped banister re-vealed a wonderful maple wood with tiger stripes. A period hand-woven wool runner was installed on the stairs. The total effect is simple elegance.

Since we moved the kitchen back into the center of the house, we were able to turn

one of the front rooms into a formal dining room. The other front room became the formal parlor. The formal dining room wallpaper is patterned with a gorgeous gold damask design. It works beautifully with the cranberry accents that were worked into the room. The drapes are a simpler swag and cascade—with the cascade hanging almost to three quarters of the length of the window. One single cranberry-colored, six-inch tassel hangs at the center of the valance. The trim is also cranberry in color to accent the gold drapes, which are self-lined. The formal shield-back mahogany chair seats are covered in deep cranberry velvet fabric and a pattern is applied with gold-headed tacks. The table they are placed beside is a three-pedestal Duncan Phyfe style in tiger mahogany. The wool carpet beneath is a cranberry color with small fleur-de-lis medallions surrounded by a border of gold garland. An early nineteenth-century mahogany sideboard stands beneath a George III convex bulls-eye mirror framed in gilt wood. A Horace Tift presentation banjo clock keeps time mounted on the wall between the two front windows.

Front foyer-backside (left)   **Dining room (below)**

Dining room

The formal parlor wallpaper is another fabulous toile but with very fancy detail: sort of Arabesque in design. The colors are striking in green, gold, white, and tan over a coral background. There are cherubs, birds, flowers, oval patterns, and draping garlands throughout. The section beneath the chair rail has the appearance of wainscoting. It has the complementary colors of gold and tan in square and rectangular shapes. The draperies are the same style as those found in the formal dining room. The material is more formal with two-tone stripes running vertically. The trim is green to match the green in the wallpaper. The tassels are gold with accents on them. The wool rug in this room is simply a work of art. The base color is a neutral grayish tan on which there are small octagon shapes outlined in black with gold layered within and coral in the very center. The black outside border is about six inches wide, with medallions of gold and coral in the center surrounded by gold borders.

To furnish this room in period reproduction pieces a trip to Boston, Massachusetts, was scheduled. There we visited a design center that could outfit any home in high style. We selected a Sheraton couch made

Formal Parlor

with carved wood trim on the upper back section and curved mahogany arms above wood legs. We chose a gorgeous gold damask fabric to cover the seat and back. We also ordered two lolling chairs with high backs and curved mahogany arms that flowed into wood legs. We had hoped for a fabric with bold vertical stripes but were unsuccessful in that endeavor because of our time constraints. The coral we needed to match in the wallpaper proved to be difficult to come by in a matching fabric. We compromised with another gold fabric with a subtle pattern of small diamonds through which two-toned lines run vertically. We also ordered three chairs upholstered in silk with green and gold stripes. The two side chairs flank a fabulous antique mahogany library table. Upon this rests photographs of our forefathers. They seem to fit there so perfectly. The third armchair was placed alongside a period mahogany secretary that Tom and I found in New Hampshire.

Formal Parlor

Kitchen with fireplace

The most popular room in the house ended up being the kitchen. We ordered a custom-made, traditional farm-style table with recycled pine floorboards for the tabletop.

The antique boards were stained a warm medium brown and a wax finish was applied over that. The table is supported with carved, turned legs painted black. The chairs are a black high back Windsor style. Under the table is a floor cloth with a diamond pattern in burgundy, cream, and black. The window treatments and chair pads are all made with the same checkered burgundy and mustard fabric. The walls are painted a maize tone.

The mustard wainscoting matches the wallboards on the center fireplace. For the stove, we chose a gas top with the heavy black burners to evoke an antique appearance. The pantry off the kitchen was reconstructed from one originally found in Canada. It has three open shelves stocked with a various assortment of antique boxes, firkins, butter presses, cranberry picker, all shapes and sizes of pottery, crocks, wooden measuring cups, bowls, mortar and pestle, and an apothecary with numerous drawers purchased at an auction. Beneath the shelves is the base with drawers and cupboards and an opening holding salt pork containers, jugs, and other pieces of stoneware. Under the stairs are the original cupboards. These are painted a country green and red in simulated milk paints. An antique dough-box is located on the adjacent wall.

Last but not least, the second most lived-in room in the house is the den; also known

Kitchen fireplace (below)

Pantry (right)

as the keeping room. This is located in what had been the kitchen when we first walked through the house. The green and cream wallpaper is a different kind of toile that has horizontal ovals depicting different scenes surrounded by slim branches. The wood trim, doors, fireplace cupboards, and wainscoting are all painted green with milk paint. The couches selected are a country look in a fabric complementary to the wallpaper. The wood pieces are reproductions made at a local furniture store. The cherry coffee and end tables are hand-planed with mortise-and-tenon construction. The Shaker-style entertainment center is made of solid cherry. It was delivered in two pieces because of its weight. The glass used in the doors is made with deliberate imperfections to look hand-blown. On the mantle sits an Eli Terry scroll and pillar mantle clock made in Plymouth, Connecticut.

Keeping room

Initial Inspections of the Barns | Chapter 12

$\mathcal{T}$om was fortunate enough to meet one of the prior owners of the property, Walter Thorp, who was ninety-six years old at the time. In his youth, Walter had done a project about the farm while studying at the University of Connecticut. At that time, in 1926, he took a photo of the farm (see page 25.) He eventually purchased the farm in 1946, but was then forced to sell it in the early fifties when his health declined from trying to run both the farm and administer the local office of the newly formed Federal Milk Subsidy Program. Tom spent quality time interviewing Walter Thorp, whom gave him much information about the property. For instance, there came a period in time where it was mandated that milking-cows could no longer be housed on wood if one planned to sell the milk publicly. Thus, the fifteen-foot stone apron at the end of the large barn was removed along with all the soil at the end of the barn. This would open up the basement, upon which the concrete floor was poured. Mr. Thorp directed us to where the stone had been relocated, which happened to be at the Catholic Cemetery where it was being used as a bench. We were able to acquire it from the church.

Walter Thorp with his prize cow (left)

Basement of barn before (below)

My first look at the interior of the barns occurred on the same day that Tom and I planned on walking the land, including the woods. Similar to the house, the basement of the barns would also need a considerable amount of work. The original posts that supported the structure had been removed and replaced with steel columns. Some of these had rusted out and collapsed over the years; others could be easily pushed over. We concluded that the large beams must have been removed to allow for a center aisle for the cows to walk down and into the double row of stanchions.

The wood had been whitewashed and much of that was peeling off. A deteriorated wood frame held the broken windows

along the southwest exterior wall, under which were concrete blocks. The concrete floor had trenches for an automatic manure system that had not been in use for many years. The stanchions made of metal were also corroded with rust and held up with the aid of the concrete floor. There were some smaller wooden pens thought to be for young calves on the street side of the basement. The foundation of the barn basement was constructed with hand-laid stone. There were a few windows that needed to be replaced.

Upstairs on the first floor, the barn was dark and gloomy, with just a few windows shedding some light. Most of the original windows on the southeast side were either boarded up or missing and added to the darkness of the interior.

There were three full-length windows in the front of the barn but these did not supply any light into the main area of the barn, as they were in separate rooms. One was located in the entry, another in what was originally an office, and the third was what looked to be a utility room that abutted the office. On the street side of the barn, the original boxed frames for small windows were also boarded up and sided over. The huge doors at both ends of the barn were hung on sliders, which needed repairing— as did the doors themselves. The windows in the doors and most of the trim-boards on them needed to be replaced as well as part of the frame and exterior siding. On the

Backview of large barn

southwest side of the barn, the original manure trenches, where the wood stanchions had been, were still intact but pretty much rotted out from cow urine. On the opposite side was a large open area with a primitive set of stairs leading up to the hayloft. The wood flooring on the main floor was the original thick, wide-board chestnut planks; some in decent condition. Other areas had

Large barn center aisle

been replaced with mismatched boards. The wall siding was a five-inch, tongue-and-groove pine board that had darkened to a tan gray color. We needed to have a large quantity of these boards milled to match. Going up the treacherous makeshift stairs to the hayloft was somewhat scary. Once upstairs, I could get a sense of what Tom had seen on that first day. The huge, expansive space had high ceilings vault-ing up toward infinity. The chestnut roof rafters at forty-six feet long stretched to the peak and disappeared into darkness. The dimensions of the large two-story barn are forty-six feet wide by one-hundred-and-four feet long by forty-six-feet high, includ-ing the full walkout basement beneath. The hayloft capacity is approximately ten thou-sand bales.

Manure trenches

Upon further inspection, we noticed that the eight-inch square beams that tie the knee brackets from the front section of the roof to the back section (acting as collar ties) had been cut out. These beams, which had been removed, were about sixteen feet long or approximately the width of the downstairs center aisle. This had been done sometime in the past to allow a make-shift hay cart to roll down the center of the hayloft. This resulted in the sides of the barn falling outward and the roof joint spreading out and falling down about twelve inches. To repair this damage, chains were placed around each beam at fifteen-foot sections and attached to come-a-longs, which were cranked inward. This needed to be done in a slow fashion. It took about two months to pull the barn back to its original position. The scarf joints all closed up perfectly, as did the sidewalls that were pulled in about nine inches. The loud creaking that went on during this process would send all but the bravest running. Once the barn was squared back up, the beams that had been cut out were replaced. Upon further investigation it was also discovered that nearly all joists and decking of the hayloft had been removed. All but a small section on the right hand side of the barn were missing.

The smaller barn with a walkout basement measured thirty-feet wide by sixty-feet long by forty-two feet in height. It was similar in condition to the larger barn. Along the right side of the main aisle was a stall where young calves had been kept. The floors rotted out from excrement and urine.

In the corner of the barn closest to the house there was a walkway about fifteen feet wide that was originally used for the cows to walk from the ground level up into the barn.

Hayloft in large barn

Rafters in large barn (below)

Basement walkway for cows in small barn

This walkway was made of large, uneven stones—not designed for human traffic—that wound upward toward the first floor. On the left side of the aisle there was decayed post and beam frames and flooring. All of the floor joists and flooring for the hayloft had been removed, leaving only marked evidence that they had ever existed. This barn needed to be pulled in as well. It shrieked even louder than the first, as it had expanded outward even further. Down in the basement we discovered large tree trunks as the support beams located only near the stone cow walkway. The rest of the support beams were missing, which had allowed the whole center of the barn to cave in about fourteen inches. The stone foundation was on three walls with the gable end reaching twelve feet in height; it was finished in cut granite stone on the exterior. The wall facing out toward the back holding pen was cement block on the bottom. This supported a row of broken-down window frames housing windows in the same dilapidated condition.

Where the two barns conjoined the roof had leaked, causing much damage to that corner. We needed to replace major beams in that area as well as some of the roof itself.

The windows and their frames on both sidewalls were in bad shape and needed to be repaired and/or replaced. In addition to the obvious need for a new roof and siding, like the house, there was much work yet to be uncovered.

Water damage where barns conjoin

Rear view of barns

The Restoration of the Barns

*W*e actually started the barns a short while before the house using two separate crews. One of the crews had worked on post-and-beam construction previously, hence the reason they were hired. Shortly, they were instructing our modern framers in the old technique since both buildings required attention at the same time. For the modern masters, this was initially a culture shock. They would normally frame a moderate sized house in a few weeks. Working with these large beams in a mortise-and-tenon method was as slow as molasses in comparison. As time wore on, camaraderie developed between the men, and the job moved along, albeit at a slow pace. We had entered into the early nineteenth century, where time moved at a different pace and nothing could be rushed. Much of the work would need to be done by hand and the materials were mammoth.

As with the house, we started in the basement of the barns. As mentioned earlier, the support beams were either missing or inadequate. At a local lumber mill, we had thirty, twelve-inch square, white oak timber beams milled to size from our own trees.Each timber weighed three hundred to

Camaraderie

four hundred pounds. As one can imagine, the men dealing with this kind of weight were no one-hundred-pound weaklings! After these were set in place, we needed to repair some of the floor joists and the underlayment above them. To get to these problem areas, first we needed to remove the chestnut flooring. Once the chestnut flooring was removed it was taken out back and power-washed to remove the hundred

years of accumulated filth and stench. It was stacked and air-dried in the sun. The next step was to remove all of the three-quarter-inch underlayment and replace it with one-and-a-half-inch white oak. We didn't think that the antique chestnut would be strong enough to hold the weight of multiple twelve-hundred-pound horses. Next, the three-by-eight-inch floor joists in the rotted areas were removed and discarded.

We would go through thirty thirty-yard dumpsters just from the barn, its attached sheds, milk room, etc., over the five-year period that the restoration project lasted. The other areas that had been affected were below the sliding doors on the north end of the large barn and the east end of the small barn. Over the years water had seeped under both doors and caused damage. Luckily, this affected only the first fifteen feet or so into the barn lengths. From about halfway down the large barn, where

Finished basement in large barn

**Barn subfloor-
ing replaced**

the animals had been originally housed on the right side, nearly all of the flooring needed to be replaced. More decay was found halfway up on the left exterior wall— it was caused by a leak from the dormer above. The bottom section of the roof rafter on which the dormer sat had to be replaced due to the water damage. The leak was remedied when the new roof was re-placed. The rotted timber below was twenty-two feet tall by twelve inches square and needed to be replaced. This was done in the same manner as the smaller beams on the corners of the house, as previously described. Another large beam at the end of the barn, where the milk room had been added, needed to be replaced as well as some of the sheathing in that area.

**Large barn
decay under
dormer**

Next we addressed the gloomy interior of the large barn. The original structure only had three windows on the front, giving it a boxed appearance. Since we didn't think this had much curb appeal and we wanted to improve the natural lighting and ventilation, we decided to add windows along the whole front of the barn. We chose windows that matched the windows on the north side of the small barn. We replaced some of the frames on those already located on the south side and installed the new six-pane window sashes. We also replaced the windows in the hayloft and on the large sliding doors.

At the same time, the repaired floor joists and underlayment were now ready for the wide chestnut boards to be re-installed. These would maintain the antique look of the interior of the barn. Since sections of these were missing, we needed to find additional matching boards. A lead to a person that disassembles barns took us again to Massachusetts, where we found the chestnut boards from two barns that had been dismantled. There would only be enough for the large barn floors. For the small barn floors, we found antique pine boards nearly three inches thick that came out of the attic of an old house, where for many years cheese had been manufactured.

In addition to adding the windows on the front of the barn, we also added another doorway with two four-foot doors that

Corner of barn with rotted beam

would swing inward into a fifteen-foot aisle.

The windows and trim were copied from those on the large sliding doors. We obtained another large stone for the apron going out of this doorway.

Next our attention was directed to replacing the floors of the hayloft. Prior to that, a set of steps leading up to it was built above the existing steps going down to the basement. After that was finished, we retrofitted the first floor with horse stalls. Building interior walls at each main beam section created the stall sides.

For the front sliding door on the stalls,

custom-made solid iron grills finished with a powder-coated dark brown color, for the aged effect, were placed above an angle-designed lower wood base. These lower sections of doors are made up of two layers of two-by-six-inch tongue-and-groove pine running diagonally on the face (to match the trim boards on the large sliding doors at the gable ends of the barns) and horizontally on the reverse side. The panels on each side of the center door are also double-layered with the pine running vertically. Custom heavy-duty track was procured to support the weight of these doors. The stalls ended up being approximately fifteen feet square in dimension.

On the floors of the stalls, to keep urine from dripping down through to the basement storage area, a one-piece, sealed and cushioned rubber mat system was glued down. Since this barn would be used to house our broodmares and deliver their foals, a high-tech monitoring system was installed, which can be viewed from inside the house.

After that segment of the large barn was done, bathrooms were constructed in the area near the alley where the barns meet. For the front entryway walls of the barn, bead-boards were copied from the originals. Because the barns were built in the Victorian age, the trim work is completely

Finished interior of barn

different than anything found in the house. The bead-boards are made of pine and are three inches wide with a bead at one edge and tongue-and-groove on the other. These boards were also used in the office and utility room. In addition to the bead boards used for the wainscoting in the office a custom-made, built-in bookcase with lower cupboards was constructed. On the floors of these three rooms, wide antique pine or oak boards were installed. The masons stopped working inside the house long enough to plaster the ceilings of both of these rooms and the top section above the wainscoting in the office.

The exact replicas of the wider tongue-and-groove pine boards without beading were used on the rest of the interior walls throughout both the small and large barns. All of the newly applied wall finishes were stained to match the existing grayish tan color, in keeping with the antique appearance.

From here we drew our attention to the small barn. It needed substantial sheathing replacement, especial-

ly at the gable end and the section above the basement windows on the southwest side of the barn.

Some of the window frames were rebuilt and new six-paned windows were installed. The rotted floorboards on both sides of the main aisle were removed and replaced. The earlier mentioned "cheese" pine floors

Gable end of small barn

Gable end of small barn—sheathing replaced

were then laid over the top of the oak sub-flooring. The missing post-and-beam framing on the first floor ceiling and the hayloft above were restored. Horse stalls were also built in this barn, but smaller in size. Instead of iron grills at the top section of the doors, these are half doors that are open at the top and swing out (instead of sliding open).

A set of doublewide stairs from the basement to the first floor replaced the stone cow walkway. A railing was added for safety as well as to meet the building code.

House wrap, which is a wind protector, was applied to the exterior of both barns. In the meantime, the multi-tasking masons closed off the basement entrance at the end of the large barn that was originally made for the dairy cows, and they fixed the stone basement window wells. I also insisted that, for function and safety reasons, we fill in the area at the end of the small barn that dropped off ten feet. This required hiding the beautiful stone foundation. Tom had a real problem with this as did I, but in my mind it was neces-

Wind protector wrap

sary. We would need more than one way to exit the barn with the horses. Of course, heavy equipment was needed to fill in that large area. Additionally, the road alongside the small barn would need to be elevated about seven feet to make the transition feasible. Because the grade was so steep, it required a gradual slope out toward the fields. Afterward, the masons hand-built two retaining walls where the fill had been added. These were made of cut granite, which helped ease the pain of covering the stone wall foundation.

At this time, the basement wall—a bank of windows over cement blocks—was re-placed with traditional sliding barn doors. These doors are sixteen feet long by eight feet high, sided with three-inch cedar boards. The doors are topped with six-pane windows to match the exterior doors used throughout the barns.

While all of this other work was taking place, another crew was working on the roof. The old rusted corrugated roof was removed, which revealed the original cedar shingles.

It took a concerted effort to remove these, as they were installed with cut nails. A crow bar was used to separate them from the oak boards. It took four weeks just to strip

Roof with old cedar shingles

Roof stripped

the existing shingles. After they were finally removed, we had two choices: add sheathing to enable us to use modern shingles or purchase Southern yellow pine wood shingles. This latter way would be the harder road, but the crew insisted they were up for the task. The installation took another four months and required two tractor-trailer loads of the wood shingles. The shingles are each three-quarters of an inch thick by twenty-four inches long as opposed to the finer ones used on the house that are a half-inch thick by eighteen inches long. In proportion they look exactly right on the huge barns.

Another area that needed work was the elaborate roof rafter tails. Each of these is carved with a unique design instead of the usual straight end cut. This gives the effect of decorative trim under the roof overhang. Some of these broke off and were missing. We made reproductions and replaced them along the length of both barns. We ordered scalloped cedar shakes to install at the peaks on both ends of the barns and on the front dormer. These added charm to the Victorian-era barns.

When removing the original clapboard siding, we noticed the company name of Granville Manufacturing Co., Inc., stamped on the backside of the siding. We contacted the company in Granville, Vermont, and discovered they were still in business! They were still using the same machines

Interior view of roof stripped

Roof shingles installation

to manufacture the siding, but those were now run on electricity instead of the original waterpower. The siding was available in random lengths of quarter-sawn Douglas fir. Before its delivery, the siding was first primed and then painted with one coat of barn-red paint. The second coat of paint was applied on site.

In the meantime, Tom had been working on having cupolas made to match the original ones shown in the photo given to us by Walter Thorp. We could not duplicate the original size of the one on the large barn, as it was too enormous to replicate off site. Additionally, when the center front door under the roof dormer was added, it nearly split the visual length of the barn into two equal lengths. By making two cupolas, the visual effect would be more pleasing. Coincidentally their combined footprint is almost the same as the original single cupola. The finished sizes of each of the two cupolas on the large barn are nineteen feet tall and seven feet at the base. The one on the smaller barn is actually bigger at twenty-one feet tall and eight feet at the base. This one is actually the same size as the original cupola on that barn.

The cupolas were delivered on a tractor-trailer and took up nearly the entire bed. Standing next to them we were small in comparison. They were exactly what we had envisioned. We needed to hire a crane operator to have them set in place. The crew up on the top of the two-story barn, forty-six feet in the air, looked like children in proportion. The crane set all three cupolas in one day. A stream of local residents was gathered on both Bunker Hill and South Street to witness the final crowning of the barns.

To top it all off, one-hundred-and-fifty-watt, heavy-duty service bulbs were installed inside each cupola to serve as night-lights for the historic homestead.

Setting cupola

Close up of cupola and dormer

The Fields

Chapter 14

The fields, as previously mentioned, were in dire need of attention. Along with the brush, there was also scrub trees and the like that needed to be removed. The mulberry rose bushes were pulled out by their roots. This was done by wrapping chains around their base and then attaching the chains to a tractor. There were literally hundreds of them that needed to be torn out. It took weeks and a determined individual to complete this task.

The plowing, harrowing, and seeding of the fields took nearly two seasons. This was done in sections all the way down to the very bottom fields along Hop River Road. It proved to be a very difficult job. When last plowed, it had been abandoned and left in furrows. This created ruts and two-sided sod that was hard to break up and level off. Additionally, livestock had been pastured, creating even more divots from their hooves. We hired a neighboring farmer, D. J. Fowler, who had a commercial tractor and equipment to do the plowing and harrowing. A landscaping contractor, Anthony Fratianni, did the stone picking, seeding, fertilizing, and pesticide application with a smaller tractor. He also signed on to perform a myriad of other jobs around the farm.

Fields being plowed

Fields made ready for seeding

**More work
on the fields
(left)**

**Fields in
rows to
be baled
(below)**

The Arena & Pond

*A*s negotiations between Tom and I continued, the pond was an inevitable topic of discussion. I brought to Tom's attention that the pond at our previous residence allowed me to swim with our Labradors. I suggested to Tom; "I couldn't possibly move to a place without a pond." Tom pointed out that there were a few "ponds" already on the property. But a watering hole is not considered a pond. Regardless, because of the town's strict wetlands regulations we weren't wholly optimistic about obtaining the necessary permits to construct a pond. We needed to have a public town meeting and obtain approvals to get permission to proceed. The meeting was scheduled to take place at the meetinghouse in town. This facility turned out to be too small to hold all of the people that attended. The meeting place was moved to a larger room in the Town Hall. Of course, Tom came well prepared with his engineer's reports, facts and figures, size, location, etc. He did an outstanding presentation, as did the other professionals who came to speak on his behalf. All of the questions put to them by the officials were answered. Many of the townspeople came to speak on our behalf as well.

Eventually, it would be voted upon but not that night. Ultimately, Tom was successful in gaining acceptance and getting his approvals.

The pond is two and a half acres and at the deepest point it is twenty-one feet. Its depth allows us to stock pretty much any freshwater fish we choose. In my mind this is an added bonus. We dug the pond during the coldest winter months in order to not disturb any wetlands. As luck would have it, that winter was the coldest it had been on record for many years. The temperature dropped to under freezing for the duration of the dig, which was one full month. Every morning it was a major job to start the equipment. The cold froze anything liquid, including the diesel lines. The contractors needed to bleed the lines every night. Another big problem associated with the cold was the dirt that lodged inside the tracks of the equipment. This needed to be removed each night so that it wouldn't freeze onto the tracks. Again, the right group of committed individuals got the job done. It would take until spring to really establish the pond. In total, thirty-three thousand cubic yards of material was dug out and stacked on the field along Hop River Road.

View of pond with Tom riding horse in the distance

This fill from the pond spoils was eventually used in raising the grade along the side and end of the small barn and to gradually slope out to the fields. Additionally, the fill was used to elevate the area upon which the riding arena was built. This was necessary because when viewing the arena from Bunker Hill, we didn't want the tall antique barns towering over it. We also minimized the grade change by building the arena roof using a double wood truss system. This method allowed for a steeper roof pitch than a more traditional steel arena. The arena took another eight months to build. We included attached stalls to stable the show horses in training. The two miles of thoughtfully designed horse fencing made for a smooth working facility. The outdoor sheds work perfectly for the horses' turnout and keep them protected from the elements. We stained the arena close to the color of the pressure-treated pine, which is a grayish tan. We chose this color for two reasons: first, we didn't want to paint the horse fencing white, as we thought it would draw the eye to it instead of the panoramic view; second, we didn't want the arena to compete with the red historical barns. We thought that since the barns are the main focal point of the farm they should stand out.

Tom cantering at base of horse fencing

The Formal Fence

*S*ome time during the project, in an effort to distinguish the house proper from the barn commons, we needed to establish the location of our fencing. We also decided we needed a second driveway to the barns and the horse arena. This would enable us to have a separate formal entrance to the house. First we determined the sideline on the north side of the house. This would be where the large paddocks along South Street would be located. In order to visually create a straight line for the horse fence, we had to minimize the size of the side yard. We settled on putting the horse fence approximately sixty feet from the house. The corner of this three-rail wood horse fence ties into the white formal house fence that juts out closer to the road. The formal fence runs along South Street in front of the house to two stone pillars that hold automatic gates. The stone pillars were erected by the masons and topped with period lighting fixtures. The stone pillars were built to match the foundations of the carriage shed and antique barns. The masons also built a retaining wall made of fieldstones along the side of the driveway, running alongside the porch side of the yard. Stone steps lead to the kitchen's

Sugar maple tree

double French door. In front of the three steps a pair of horse heads is mounted on vertical granite stones and black wrought-iron chains attach them to the stone wall. At this point, the width of the driveway is reduced, as there is no longer a need for trucks and trailers to utilize it. From the stone pillars the formal fence continues to run past the wonderful huge old maple tree to the second driveway, which has access to the barns and arena.

The lovely white formal fence has a hint of Victorian flare. It consists of eight-foot sections made of cedar. At the end of each section there is an eight-inch square post with a recessed, carved design upon which sits a decorative cap. Each section is made of two-by-two-inch square stiles or pickets three inches apart that are enclosed within two-by-four-inch rails, instead of the more common face-applied pickets. These stiles are level in height until they get into close proximity to the posts, at which time they gradually increase in height. The posts at the front center gate leading to the formal front portico are slightly larger at ten inches square. One other unique and complicated feature is the curved angles that tie into the stone pillars. These have a twenty-two degree, rounded compound curve. These rounded sections were custom-made out of cedar at a woodshop in Long Island, New York.

House with close-up of compound curve

Landscaping

*A*nother major decision we worried over was whether or not to plant a row of trees along South Street. We were afraid to block the spectacular view with its sweeping vistas. We made our decision by measuring the entire length that would be affected, putting in stakes approximately where the trees would be planted, and then driving by to see how it looked. By spacing them far enough apart, we determined that there would be minimal negligible affect and a major decorative improvement overall. We wanted a tree that wouldn't end up growing wide at the top and that wouldn't be poisonous to our horses. We decided on a tree called the Zelkova Serrata Halka. It was developed to resemble the old elm trees that were used years ago along streets and in parks.

Ten of these trees were planted on South Street and a few more in the barnyard and inside the paddocks to offer some shade. Each tree weighed approximately twelve thousand pounds. They were eleven inches in diameter and twenty-five feet tall. They were delivered two per tractor-trailer truck. Two other trees, a European Beech and Copper Beech, were planted in the yard area on the north

Setting trees

side of the house. These were even larger than the others at fifteen thousand pounds, fourteen inches in diameter and thirty feet tall. Each tree was delivered independently on a tractor-trailer. All of these trees had to be lifted with equipment designed for the job and set into the prepared holes.

In addition to these trees, we acquired a gorgeous, perfect specimen Fraser fir (that had never required trimming) from a local tree farmer. It was planted on the lawn near one of the compound curve sections along the formal fence. We also planted two other large evergreen trees near the large barn entrance as well as forty others throughout the barn and arena commons. Two tall specimen trees, Serbian spruce,

Row of Zelkova Serrata Halka trees

Back of barns

were also planted near the tall barns. Four Weeping Sergeant Hemlocks were planted along the house driveway as a buffer between the residence and barnyard; more were planted on the area that had been filled near the end of the small barn. Flow-ering trees were also added to the land-scape, including dogwood and pear. A sweet-smelling bush was planted just out-side the entrance of the barn as well as the more traditional lilac.

Front entry gate

Last but not least, a landscape designer drafted the finished formal house landscaping plans. The plans depicting a round parterre circling the front gate to the portico entrance earned him the job. Inside the parterre are purple perennial flowering plants with a white peony in the center. When in full bloom, this is spectacular. The best view of it is from the second story window in the upstairs foyer. Many other period shrubs and flowering perennials designed to bloom at different times of the season were planted around the house proper and along the formal white fence. In keeping with the period of the house, care was taken to avoid planting any shrubs close to the foundation with the formal cut stones. Another parterre was created to surround an antique fountain with Pan, which I had purchased a few years earlier. He stands right outside the kitchen window where I admire him daily. The base that surrounds the statue is made of the same reddish brick that was used on the formal entrance walkway, and is close to the color of the chimneys. Both parterres are circled with boxwood shrubs as well as the stonewall perimeter running alongside the porch side yard next to the driveway.

View of mare with foal and the statue of Pan (above)   **Porch view 2 (below)**

Holly berry bushes in the form of shaped trees flank the formal portico. Other more natural shaped holly berry bushes are randomly placed in both the house and barn commons. The red berries and green leaves up against the red barn look exquisite in the wintertime. Of course, a Connecticut landscape wouldn't be complete without mountain laurel. Two of these are planted at the very corners of the front of the house balancing it nicely.

All of this gorgeous landscaping transformed what was originally a commercial looking, wide-open space into a residence with the curb appeal that we had envisioned.

Autumn

Home Sweet Home

*T*he move to our new residence is behind us and we are now all settled in comfortably. The large broodmares, heavy in foal, enjoy their new spacious accommodations within the majestic antique barns. The stalls are actually large enough so that within a few hours of birth the foals are already running around their dams. The hay stored in the lofts above them produces the sweetest smell. In the summertime, even during the hottest of days, a constant breeze flows through the windows keeping the barns cool. Another unexpected pleasure is the sound of the horse's hooves atop the creaking old floorboards, which produce a heavy yet hollow clopping sound.

The four cats keep the barns free of unwanted rodents. The Labradors, contained within the borders by invisible fence, have the run of the place. They are the self-imposed guardians at the gate. In the barns, where they sleep, no one dares enter without permission. They enjoy fetching sticks that we toss into the pond. They make feeble attempts at chasing the Canadian Geese that oftentimes stop by for a rest. There are families of ducks that paddle off when approached. Even a few long-legged cranes standing along the shores have been spotted. The large-mouth bass have become quite plentiful and on occasion can be seen jumping out of the water. There are so many that we guarantee family members with children that a fish will be caught on nearly every cast. The walk through the woods to the pond is an adventure all in itself. Some of the youngsters visiting from the city have never heard a frog croak nor seen a firefly light up in the dark. They are amazed at the cacophony of sounds elicited from the many wild creatures along the way.

The paddocks, seeded with a mix of Timothy and orchard grass, provide the perfect nutrients for the grazing horses. The youngsters seem to love the open spacious fields in which to romp and play. In the spring, many a car will pull off the road just to witness the sight of a mare with her newborn foal. The rest of the reclaimed fields produce enough hay to feed the herd throughout the winter months. Eventually, when these fields are at optimum production, we hope to sell the extra hay as a source of revenue.

Tom is, of course, thrilled to be residing here. On occasion, he rides his stallion outside the arena confines. The length of Bunker Hill makes for a long canter back up toward the barns. I'm certain he reminisces about having done the same thing as a child riding his pony. The office in the barn comes in handy. It allows him to

work from home on his "real" job most of the time. He is pleased that our planning on the layout of the farm has evolved into such an efficient working facility.

Occasionally, when the farm work is done, Tom and I relax on the porch in the old white rockers and breathe a sigh of relief. We are happy that the restoration is complete and pleased with the results. We hope to be around to enjoy the place for many years to come. We are grateful to those who came before us that the property was never subdivided and that the farm was kept intact.

Barn restored

House

The house, which is now our home, is very comfortable. It offers a feeling of warmth and solidity. I oftentimes just walk around and marvel at all the details. I can appreciate all of the work that was done having witnessed the progression.

Sweeping vista

Sometimes, I feel more the caretaker than the mistress of the house. I wonder, after all, if it was fate that chose us. At night, when the skies are clear, the stars seem to twinkle brighter—as, in fact, we have moved closer to heaven.

Summary of Genealogy

Researched by: Anthony Burke

Introduction:

$\mathcal{T}$he house construction, by Phineas Post, began in 1809 and was completed in 1815 for the Rev. Chauncey Booth and his wife Laura Farnam. Henry F. Dimock, who was the grandson of Chauncey and Laura, built the barns in 1899. In 1911, Henry F. Dimock bequeathed to the South Coventry Library Association the sum of forty thousand dollars if the library changed its name "in perpetuation of the memory of my grandfather the Rev. Chauncey Booth, long pastor of the Congregational Church in South Coventry, and of my father Dr. Timothy Dimock, who was born in South Coventry and there practiced as a physician during his long and useful life." The South Coventry Library Association became The Booth & Dimock Memorial Library on May 20, 1911.

Genealogy

On September 20, 1815, Chauncey Booth* was ordained minister of the First Church of Coventry. Very soon after, on October 26, he bought land with buildings from Phineas Post for $2,200. Joseph Dow owned the homestead that abutted the property Mr. Booth bought. Less than a month later, on November 16, Reverend Chauncey Booth married Laura, daughter of Peter and Sylvia Farnam of Salisbury, Connecticut. The need for a house before winter was paramount, as he had a new job that required a place for entertaining and he was starting a family.

In the 1820 US Census for Connecticut, Tolland County, Coventry, Chauncey Booth was living at a house that had Joseph Dow and Ruel Loomis as close neighbors.

In the 1830 and 1840 US Census for Connecticut, Tolland County, Coventry, Chauncey Booth was living at a house that had Joseph Dow and Ruel Loomis as close neighbors. More names are repeated as neighbors from census to census. Also, there is a woman, age seventy to eighty years, listed in the household.

In the 1850 US Census for Connecticut, Tolland County, Coventry, Chauncey Booth and Sylvia are living in the same house. Joseph Dow was still a close neighbor and John Ellis was another.

In January 7, 1852, Laura F. Booth sold the property to Martin Lyman and Edward H. Dow. The land is referenced to deed entries and also has Joseph Dow as an abutter.

*Chauncey Booth, the third son of Colonel Caleb Booth, of East Windsor, Connecticut, and grandson of Caleb and Hannah (Allen) Booth, of East Windsor, was born on March 15, 1783. His mother was Anne, youngest daughter of Captain Jonathan and Hannah (Watson, Bissell) Bartlett, of East Windsor. He united with the College Church at the end of his sophomore year. At graduation he was the oldest member of the class. He studied theology for three years in the Andover Seminary, and spent some time in 1814 to 1815 in home missionary work in Bridgewater, New Hampshire.

On September 20, 1815, he was ordained and installed as pastor of the feeble Congregational Church in South Coventry, Tolland County, Connecticut where he was instrumental in building up the church and in uniting the people.

He married on November 16, 1815, to Laura, daughter of Peter and Sylvia Farnam, of Salisbury.

After a highly successful ministry, in the course of which two hundred and ninety-two persons, mostly the fruits of five seasons of revival, were added to the church, he was obliged by ill health to take a dismission on March 20, 1844. He continued to preach little after his dismission, and sank gradually under a complication of diseases.

His residence continued in Coventry, where he died on May 24, 1851, in his sixty-ninth year. A commemorative discourse by the pastor of his old church, the Rev. Charles Hyde, was afterwards published. His widow died in Ellington, in the same county, on September 6, 1875, at the age of eighty-three.

They had seven sons and three daughters. Two sons became physicians. The eldest daughter married Timothy Dimock, M.D. (Yale Medical School 1823).

Mr. Booth was a grave, dignified, prudent, and sincere clergyman, who enjoyed in an unusually high degree the confidence and esteem of his people.

Authorities. Dimock, Coventry Records, 11-12, 130, 174, 268. Hyde, Sermon at Mr. Booth's Funeral. Stiles, History of Windsor, 2nd ed., ii 112-113.

The History and Genealogies of Ancient Windsor, Connecticut; including East Windsor, South Windsor, Bloomfield, Windsor Locks, and Ellington. 1635-1891. By Henry R. Stiles, A.M. M.D. Vol. II Genealogies and Biography Hartford, Conn,.

Press of the Case, Lockwood & Brainard Company, 1892.

OBITUARY of Henry F. Dimock from the *New York Times* (Thursday, April 13, 1911):

To Rest with Forefathers: Henry F. Dimock to be buried in Coventry, Conn.,

Where Ancestors Lie.

The funeral of Henry Farnam Dimock, who died on Monday, will take place this

Booth monument

morning from the family residence, 25 East Sixtieth Street. The burial will be at Coventry, Conn., where he was born, and where the bodies of his father, Dr. Timothy Dimock; his grandfather, Capt. Daniel Dimock; his great-grandfather, Lieut. Timothy Dimock; and his great-great-grandfather, Capt. Timothy Dimock lie.

Henry Farnam Dimock was descended on his father's side from Lieut. Thomas Dimock, who settled Barnstable, Mass. The wife of the latter was Joanna Bursley, a daughter of the Rev. Joseph Hull. Among Mr. Dimock's pre-revolutionary ancestors were Gov. William Bradford, Major William St. Joseph, the Rev. James Fitch, and Major John Mason of Pequot war fame. Among

Mr. Dimock's maternal ancestors was John Howland, a Mayflower Pilgrim.

Mr. Dimock's summer home at Coventry was many years ago the combined property of his two grandfathers, the Rev. Chauncey Booth and the Rev. Daniel Dimock.

Dimock markers

Early History/Chain of Title

*T*his is a synopsis of records/outbuildings on the property known as 1209 South Street, Coventry. The 92+/- acre tract as deeded and recorded on the Land Records of the Town of Coventry from 1799 to 1938.

Tracing the property back to the early 1800's, the house parcel was deeded by Daniel Robertson to the Rev. Abiel Abbot by a Warranty Deed dated February 9, 1801; recorded March 12, 1801 in Volume 10 at Page 85—no buildings were cited. No buildings were cited in the Warranty Deed from the Rev. Abiel Abbot to Phineas Post dated April 26, 1809; recorded June 7, 1810, in Volume 12 at Page 153. Neither Rev. Abbot nor Mr. Post had any liens or mortgages recorded against the property.

Phineas Post sold the property to the Rev. Chauncey Booth by a Warranty Deed dated October 26, 1815; recorded November 7, 1815, in Volume 13 at Page 484 "with the outbuildings standing thereon." When he died, the property went to Laura F. Booth, his wife, by LW&T dated May 3, 1851; recorded in Coventry District Probate Court Volume 1 at Page 144.

Laura F. Booth sold the property to Martin Lyman and Edward H. Dow tenants in common by a Warranty Deed dated January 7, 1852; recorded January 8, 1852 in Volume 22 at Page 56. She described the property as "the farm where I now live" located on both sides of the road.

The Coventry Vital Statistics Index Volume 4 Page 7 indicates that Martin Lyman died October 17, 1859, at the age of seventy-six years, ten months and ten days. He was a farmer and was married. There is no copy of his birth certificate recorded in Coventry, there are no probate records of his estate in Coventry, and there is nothing recorded on the Land Records showing where his undivided one-half interest in this property went.Six years later, Edward H. Dow deeded the 32+/- acres to Daniel O'Brien. There are no other deeds to either Dow or O'Brien of the undivided one-half interest that had vested in Martin Lyman, and the property was subsequently deeded as though it was a 100% vested interest.

The two parcels came "under the same ownership briefly between June 1880 and April 1882." Henry F. Dimock purchased them in 1886. He purchased the 32+/- acres from Lilly W. Barney by Warranty Deed dated March 22, 1886; recorded March 31, 1886, in Volume 29 at Page 434 and the 60+/- acres from Francis C. Spaulding by Warranty Deed dated September 30, 1886; recorded October 4, 1886; in Volume 29 at page 457. They were first jointly described as one parcel containing 92+/- acres when deeded to Franklin R. Orcutt by a Fiduciary deed dated and recorded January 7, 1936.

Title Search

Compiled by William Jobbagy

Frank Orcutt 1936–1938.

Susan Catalani 1922–1936. Susan was the daughter of Henry and Susan Dimock, and lived much of the time overseas.

Susan Whitney Dimock 1911–1922. From the estate of her husband, Henry F. Dimock, who died in 1911.

Henry Farnam Dimock 1886–1911. Henry owned, over time, several large farms in Coventry.

This 92-acre site was acquired in 1886 from two sources: The sixty-plus acre tract abutting Bunker Hill Road was acquired from Francis Spaulding, who previously acquired the land (with a barn thereon) from George Marcy in 1883, who acquired it from the Lord family in 1849. This piece was in the Lord family's possession back to the very early 1700s.

The thirty-two plus acre site, on which stands the house and barn, was acquired from Lillie Barney (of New York) in 1886 and included forty acres across (NE) South Street.

In reverse order between 1886 and 1852, the owners were L. Barney, George Marcy (an architect), Loring Winchester (a financier), Daniel O'Brien, and Edward Dow. Edward Dow acquired the farm in 1852 from Laura Booth who obtained the site from the estate of her late husband, the renowned Reverend Chauncey Booth who died in 1851.

The 1850 Census of Agriculture lists the farm owned by Chauncey Booth as eighty acres (three acres unimproved) valued at $3,000 and having produced forty-five bushels of Indian corn and twenty bushels of oats that year. The farm held a horse, three milk cows, two oxen, and ten swine. Rev. Chauncey Booth acquired the farm in 1815 for $2,200 from Phineas Post. The deeds prior to this sale do not indicate any buildings on the site so it is likely the original farmhouse was built between 1809 and 1815. Mr. Post acquired the land from Reverend Abiel Abbott in 1809, and he from Daniel Robertson in 1801. The land was in the Robertson family for generations' prior and was part of the original allocation of land to the Davenport family in the early 1700s.

April 22, 1854, Martin Lyman sold his interest in the farm, "formally occupied by Rev. Chauncey Booth," to James R. Dow.

October 16, 1855, James R. Dow sold his interest in the farm, "formally occupied by Rev. Chauncey Booth," to Edward H. Dow.

The town map for Coventry published in the *Atlas of Tolland County, Connecticut*, dated 1857, has E.H. Dow as living in the house.

In the 1860 US Census for Connecticut, Tolland County, Coventry, Edward H. Dow has Hannah Dow, widow of Joseph, named as a close neighbor.

September 1, 1864, Edward H. Dow sold three tracts of land to Daniel O'Brien. One was thirty-two acres on South Street.

The town map for Coventry published in the *Atlas of Windham & Tolland County, Connecti-*cut, dated 1869, has Daniel O'Brien as living in the house.

In the 1870 US Census for Connecticut, Tolland County, Coventry, Daniel O'Brien was living in the house. The listing of neighbors on the 1869 map confirms this.

Coventry Land Records, Town Clerk's Office, Town of Coventry, 1712 Main Street, Coventry, Conn.

Notables

THE REVEREND CHAUNCEY BOOTH was born in 1783, the third son of Col. Caleb Booth of East Windsor, Connecticut. He studied theology at the Andover (Massachusetts) seminary for three years and spent some time in missionary work in Bridgewater, New Hampshire, in 1814 to 1815. In September of 1815, he was ordained as pastor of the "feeble" First Congregational Church of Coventry, then located at the site just east of the Town Green. He was instrumental in building up the church and uniting the people. He married Laura Farnam of Salisbury in 1815. He was obliged to remove himself from the ministry in 1844 due to ill health. Of his ten children, two became physicians and his eldest daughter married Timothy Dimock, a Coventry physician.

Reference:

Biographical Sketches of the Graduates of Yale College with Annals of the College History, Vol. VI, September 1805–September 1815, Franklin Bowditch Dexter, New haven, Yale University Press, 1912.

TIMOTHY DIMOCK was born in Coventry in 1800, attended the district school, and studied higher English and Greek and Latin with the Rev. Chauncey Booth. He graduated from Yale

College in 1823, was a medical student of Dr. Chauncey Burgess of Coventry, and practiced for a few years in Granby, Mass. In 1826 he married Anna Wright of Granby. In the spring of 1827 he settled in Coventry and began his practice as a physician that lasted forty-five years. In 1830 he purchased an eleven-acre tract on Cross Street from Chauncey Burgess. It was adjacent to the old school house, and at the time expanded his estate by twelve acres. Following the death of his first wife, Anna, in 1838, one year later he married Laura Farnam Booth (Rev. Chauncey's daughter). Timothy fell ill in 1872 and died in 1874.

Timothy Dimock, M.D., biography as recorded in: *Commemorative Biographical Record of Tolland and Windham Counties, Connecticut.*

HENRY FARNAM DIMOCK was born in Coventry in 1842 and died in New York City in 1911. His father was Timothy Dimock, M.D. Henry F. Dimock graduated honorably at Yale College in 1863, studied law at Harvard, was admitted to the bar in 1865, and started a law practice in New York City. For many years he summered at his family's estate on Cross Street in Coventry. He entered into law partnership with William Whitney and married Williams's sister, Susan, in 1867. Henry was director of the Knickerbockers Trust Company and the National Bank of North America. He was a member of the Governing body of NYU when the New York University Medical College was formed. In 1870 he left the law firm and took charge of the Metropolitan Steamship Company and later was appointed commissioner of docks in New York City. Henry F. Dimock left $40,000 in memory of his father, Timothy Dimock, and his grandfather, Rev. Chauncey Booth, for a new library building in Coventry. The Booth-Dimock Memorial Library was dedicated in October of 1913.

Biographical Sketches of Prominent and Representative Citizens and Many of the Early Settled Families. J.H. Beers & Co., Chicago; 1903; p. 830.

Acknowledgements & Credits

We would like to personally thank all of the individuals and companies that were a part of our restoration project, most especially the following:

A & B Lumber & Barns, LLC—Bob Austin

Advanced Building Concepts—Timothy K. Gill

Advanced Concepts LLC—Bill A. Herzog

Anthony Burke (genealogy)

Anthony Fratianni (landscape contractor)

Bakers Country Furniture—John Rossi

Booth Flooring Inc.—Brian Booth

Builders Concrete East, LLC

Contractors Forestry Services, Corp.—William Burke

Craig Shamback (timber framing)

Crossen Builders, Inc., and employees

Carlisle Wide Plank Floors

Country Classics Cupolas—Kevin Cotter

D.J. Fowler

Eleanor Racette—wallpaper

Evergreen Audio Video Communications, LLC.

Frank Racette—Antiquary

Fsaaam (inspiration)

Gale Horton (interior decorator)

Ladd and Hall Co., Inc.— Peter Olson

G.M. Bergeron, Inc.

Granville Manufacturing Co., Inc.—Jeff Fuller

Hans Hackner & Co., Inc. (painting contractor)

Hany Brothers, Inc.(plumbing contractor)

H. Ray Paine Restoration Masonry

Jeffrey A. Todd (timber framing)

Jake's Foundation, Inc.

Killingly Fences

Lagace Siding, LLC

Magic Construction—Thomas J. Crossen, III and Leslie Crossen

Meehan & Goodin PC (engineers and surveyors)

New England Saw & Lumber, LLC—Kevin E. Bantle

Nutmeg Mechanical Services

Putnam Welding—William J. Becker

Quality Paving Co, Inc.—Robert Davin

Ray Peck's Excavating-Raymond W. Peck, Jr. and Bo Peck

Sean Mullett (carpenter)

Stephen C. Breen (stone mason)

Tompkins Earthscapes, LLC—Steven Tompkins

Vermont Soapstone Inc.

Walter & Viola Thorp

William F. Bender Jr.- (wallpaper installation)

William Gould Architectural Preservation, LLC

William Jobbagy (title research)

Windham Materials, LLC

Woodbury Blacksmith

Woodworks Construction— Larry Mullett

Thanks also go to Coventry friends M. Deborah Walsh and Bruce A. Bellingham for their assistance in proofreading the manuscript.